EXPERIENCING GEOMETRY

on Plane and Sphere

Geometry

Logic can only go so far —
after that I must see-perceive-imagine.
This geometry can help.

I may reason logically thru theorem
and propositions galore,
but only what I perceive is real.

If after studying I am not changed —
if after studying I still see the same —
then all has gone for naught.

Geometry is to open up my mind
so I may see what has always
been behind
the illusions that time
and space construct.

Space isn't made of point and line
the points and lines are in the mind.
The physicists see space as curved
with particles that are quite blurred.
And, when I draw, everything is fat
there are no points and that is that.
The artists and the dreamer knows
that space is where an image grows.
For me it's a sea in which I swim
a formless sea of hope and whim.

Thru my fear of Infinity and One
I structure space to confine
my imagination away from the idea
that all is One.

But, I can from this trap escape —
I can see the geometry in which I wander
as but a structure I made to ponder.

I can dare to let go the structures
and my fears
and look beyond
to see what is always there to see.

But, to let go, I must first grab on.
Geometry is both the grabbing on
and the letting go.
It is a logical structure
and a perceived meaning —
Q.E.D.'s and "Oh! I see!"'s.
It is formal abstractions
and beautiful contraptions.
It is talking precisely about that
which we know only fuzzily.
But, in the end, and, most of all,
it is seeing-perceiving
the meaning that
I AM.

— David Henderson, 1978